It's Simple!

Money Matters
for the
Nonprofit
Board Member
Richard and
Anna Linzer

■ Port Madison Press

It's Simple! Money Matters for the Nonprofit Board Member
Copyright © 1999 Richard and Anna Linzer
Post Office Box 374
Indianola, Washington 98342
phone: 360-297-8331
fax: 360-297-8254
email: rlinzer@krl.org

ISBN 0-9669792-0-6
Library of Congress Catalog Card Number: 99-90178

Port Madison Press
9911 Shore Drive NE
Indianola, Washington 98342

This publication is designed to provide accurate and authoritative information in regard to the subject matter covered. It is sold with the understanding that it is not intended either to offer or to be a substitute for legal, accounting, or other professional services. If such advice or assistance is required, the services of a competent professional should be sought.

Copyeditor: Sherri Schultz
Designer: Sayre Coombs
Printer: Springs Printing

Printed in the U.S.A.
First Edition

Table of Contents

Introduction

People sign up to be board members of nonprofit organizations for a whole bunch of reasons. But struggling with murky financial systems or facing rapidly mounting deficits is usually not one of them. Yet, like it or not, all nonprofit board members are responsible for the financial well-being of their organization. These good-hearted souls are also called upon to interpret financial conditions and to make clear, coherent financial policies. Since these tasks involve all board members, not just those on the finance committee, fiscal management needs to be simple.

This book is for all board members, including those who may not have a clue about budgets or deficits. Using commonly asked questions, our approach is simple and direct. It starts with cash flow as a central concept and then progresses in a straightforward manner through issues concerning budgeting, financial planning, cash reserves, endowments, and the ownership of buildings.

The clarity and simplicity of this approach should offer a welcome change to board members trying to grapple with arcane notions of double-entry bookkeeping. At the heart of this book is the singular notion that all major financial issues in the nonprofit sector can be best understood in terms of the income that flows into and the expenses that flow out of organizations. By concentrating on cash flow, and then on budgeting, reporting, and monitoring, board members can dramatically improve the fiscal workings of their organization.

Definitions

■ *What is a nonprofit organization?*

The term "nonprofit organization" is a misleading term.
Nonprofit does not mean that the organization cannot earn
a profit, and in fact many nonprofit organizations enjoy
profits or at least revenues that exceed expenses. However,
in contrast to businesses, which exist to make money for
their owners, nonprofit organizations are required to use
their profits for program activities. The IRS puts it in fairly
blunt terms when it states that the prohibition on private
inurement in the nonprofit sector means that an individual
"cannot pocket the organization's funds."

The purpose of for-profit businesses is to create opportuni-
ties for private gain for owners and stockholders.
Nonprofit organizations may not do this. Economists call
this fundamental standard the nondistribution constraint.
Nonprofit organizations can provide compensation to
employees and others; however, the law does require that
these payments be reasonable. Of course, "reasonable" is a
pretty flexible term these days.

■ *Is there a difference between a board member, a director,
and a trustee?*

Not in this book, or as far as we can tell in the greater
universe of meaning.

7

Cash Flow

■ *Why can't a nonprofit be more like a business?*

This perennial favorite is a little like the question posed by Henry Higgins in *My Fair Lady* when he sighs: "Why can't a woman be more like a man?" We know there are significant differences, but some folks wish there weren't.

There are two good reasons why nonprofits cannot be like businesses. First, businesses are creatures of the marketplace; nonprofits are creatures of the IRS. After all, without a charter from the state and feds, tax-exempt organizations could not offer deductions for contributions and, therefore, avoid the taxes businesses pay. Second, the ways in which the two types of organizations are allowed to raise money are different.

Businesses sell equity (ownership) to investors, in the form of stocks, bonds, partnerships, and venture arrangements. Nonprofits are prohibited from selling equity to anyone. Instead, they are allowed to receive tax-deductible gifts and grants.

Both businesses and nonprofits can use retained earnings to pay for their activities, and both can borrow money. So the major difference is this: Businesses sell equity and distribute profits to investors and to the government in the form of taxes; nonprofits receive tax-deductible grants and gifts that are meant to be spent for the social purpose for which the organization was formed.

Businesses have a very different annual fiscal cycle than their nonprofit counterparts. Businesses mobilize money through the sale of equity, or they use retained earnings (what is left over when all the bills have been paid) or borrowing to provide the funds they need to operate. At the end of the fiscal year, they distribute earnings in the form of dividends and taxes.

The nonprofit organization is continually soliciting gifts and grants, attempting to earn revenue, and occasionally borrowing. But because there is no equity to sell, nonprofit organizations do not engage in distribution to investors and in many cases pay few, if any, federal, state, or local taxes. Therefore, the cycle for a nonprofit organization is only about cash flow: income in and expenses out.

While it is true that many business practices are highly applicable to nonprofit organizations, the two types of organizations are as different as men and women, and thereby hangs a tale.

■ *What's the bottom line in all this?*

In business there is one bottom line. Either you are fiscally solvent, or you are on the way to being out of business. At least in theory, decisions can be focused on that one bottom line, letting the chips fall where they may.

All nonprofits have two bottom lines: the immediate pursuit of their social or artistic purpose, and long-term fiscal solvency. Social purpose distinguishes nonprofit organizations from commercial entities. In this way, nonprofits are profoundly different from businesses that focus exclusively on being fiscally solvent.

In the nonprofit sector, both bottom lines command our attention, and they often compete. When the artistic director comes forth with a proposal for truly grand costumes for the show, the board may be torn between the budget and the expensive aesthetic vision of the production. Business sense might tell us to pare back the production costs, but our artistic sensibilities may support wholeheartedly the concept of making an opulent statement. This happens all the time in nonprofit organizations, particularly in cultural institutions, and demonstrates the tension between fulfilling the social or artistic mission of the organization and keeping the budget in balance.

■ *If there are two bottom lines in the nonprofit world, how the heck are we supposed to make sound financial decisions?*

Board and staff members often lack reference points within the nonprofit sector that can be used to assess fiscal policies. In the commercial world, a number of key indicators help people to see the consequences of their financial policies. To begin with, their task is easier because they have one bottom line instead of two. Beyond profitability, they can use numerous markers, including the stock market, to judge their financial decisions. Industry-wide statistics allow for comparison. Data from credit and other services help to define levels of performance. There is a wealth of information to use in comparing and contrasting fiscal performance.

Unfortunately, there is almost no useful data on financial performance available to the nonprofit sector. Attempts over the years by consultants, well-intentioned university professors, and rogue accounting firms to establish meaningful ratios and other indicators of performance have been largely unsuccessful.

■ *What do IRS statistics tell us?*

As we have noted, businesses are creatures of the marketplace, while nonprofit organizations are products of the Internal Revenue Service. And the IRS has traditionally not been particularly interested in studying its children. It will allow researchers to examine Form 990 tax forms, but the process is cumbersome. Understandably, this makes it very difficult for board members, as fiduciaries, to weigh their policies regarding nonprofit organizations. Therefore, it is difficult to apply generally held standards when trying to decide whether to go for the expensive production values or to rein in the budget.

■ *In the absence of conventional bottom-line measures, what does my board need to understand about financial management?*

We believe that all nonprofit organizations need to develop budgets that show annual cash flow on a monthly basis, not just annual totals of money spent in various categories. Foundations, corporations, and government agencies request measures of solvency such as annual budgets, audits, and balance sheets. They do this for a variety of reasons, some historical and others nonsensical. You, as a board member of a nonprofit organization, need to have a cash flow statement in your hands at each meeting.

■ *Our financial reports are confusing. Is there any way to tell the forest from the trees?*

Three tips help to make nonprofit financial reports much more user-friendly. First, look at cash flow budgets that reflect monthly totals rather than annual statements of income and expense. Second, ask that there be footnotes for every item listed in the budget to explain exactly what is meant. And third, try using a three-line budget with a line each for income, expense, and the running or cumulative total. If you convert it to a graphic, it may be even easier for everyone to read.

■ *Isn't it best to get an accountant and a lawyer on the board and let them deal with the complexity and legality of the numbers in the budget?*

The nonprofit sector is the land of hopes and dreams, where highly trained professionals sometimes run into trouble when they try to apply the skills they honed in their profession. We strongly believe that the role of the board is to help the organization address and solve complex problems. Professionals who have a fiduciary responsibility as board members need to focus on the policy-making aspects of their role. They should not be granted special authority because of their professional knowledge, which may or may not be applicable to the particular problem at hand.

There is another reason for our bias in this area. It is always nice to have an accountant or a lawyer on the board, but it makes little sense to draft these people into

12

doing chores that they perform all day long. The few dollars saved each year are hardly worth requiring people, who may want or need a change of pace, to extend their workday every board meeting. We suggest to all our clients that they employ professionals. The board member who is the lawyer or accountant or banker can discuss the terms and conditions of the assignment with their professional counterpart. But then the board needs to let the hired guns do the work.

In addition, what you want is a simple financial system that every member, even the most mathematically challenged, can understand. All board members, not just members of the finance committee, are responsible for the fiscal affairs of the agency. Strive for clarity in reporting and high marks in comprehension, and let the pros on your board have a night off.

Why don't nonprofits compete with each other the way businesses do?

Nonprofits compete big-time; they just don't brag about it. All nonprofits compete for a relatively inelastic pool of resources, but it's just a convention within the charitable sector to avoid direct references to competition; and in fact, competition in the nonprofit world is often more oblique than in the commercial sector. Still, with finite resources available each year, not only do nonprofits compete for earned revenue and access to credit, they also compete for dollars in the form of gifts and grants.

■ *Shouldn't the fiscal policies of an organization reflect the difference in size between large organizations and little organizations?*

Large sailing vessels and tiny dinghies all operate according to the same principles of sailing. The differences in size do not change the way in which the wind blows, or the need to navigate safely. So too, in the nonprofit sector, differences in the scale and complexity of institutions do not change the basic principles that are shared in common. These are important for board members to grasp, whether they sit on the board of Stanford University or the Little Two Shoes Dance Company.

All board members need to understand the financial condition of the organization. Each needs to know what strategies are available and appropriate for dealing with circumstances that arise. Board members should comprehend what is morally and ethically in bounds and out of bounds. They need to see the roles of staff and of the board in a context that enables them to provide coherence to their institution.

■ *Isn't it important for little organizations to adopt the fiscal systems of larger organizations if they are to grow and stabilize?*

Don't roll out a cannon to shoot a flea. Adopting a financial system appropriate to the level of complexity of the organization makes sense. Growth is a function of the fiscal systems only to the extent that good monitoring and efficient use of capital will help any organization to proceed. Excessive systematization of accounts, heavy-duty planning efforts, and massive budgeting functions burden and may retard the development of smaller, simpler institutions.

■ *What fiscal responsibility do we, as board members, have?*

The trustees of a nonprofit organization are asked to look out for the care and feeding of the agency on behalf of the public. Board members are meant to oversee the financial health of the organization; to take the necessary steps to see that adequate resources are in place to fulfill the mission; and to determine that good practice has been followed in the securing and spending of funds.

To our way of thinking, this means that directors are responsible and, in some cases, liable for the deeds and misdeeds of the institution. Insurance is available to directors and officers. Some states allow for limited in-demnification. Still, prudence is the virtue that most successfully reduces risk.

Forecasting

■ *What does a cash flow budget look like?*

Sample Cash Flow Budget

Income	Jan	Feb	Mar	Apr	May	Jun
1. Grants	0	5,000	15,000	20,000	30,000	0
2. Donations	5,000	10,000	5,000	2,000	5,000	2,000
3. Gifts	4,000	4,000	0	7,000	2,000	3,000
4. Earned Revenue	21,000	13,000	18,000	16,000	13,000	35,000
Total Income	30,000	32,000	38,000	45,000	50,000	40,000
Expense						
5. Salaries	20,416	20,416	20,416	20,416	20,416	20,416
6. Fringe Benefits	4,491	4,491	4,491	4,491	4,491	4,491
7. Part-time Wages	5,000	8,000	2,000	0	0	0
8. Consulting Fees	0	0	0	0	0	0
9. Rent	2,000	2,000	2,000	2,000	2,000	2,000
10. Utilities	500	500	500	500	500	500
11. Legal/Acctg	3,500	2,000	2,500	500	500	500
12. Supplies	2,093	1,593	1,093	1,093	593	1,093
13. Printing	3,000	1,000	2,000	3,000	1,500	6,000
Total Expense	41,000	40,000	35,000	32,000	30,000	35,000
Running Total	(11,000)	(19,000)	(16,000)	(3,000)	17,000	22,000

16

Jul	Aug	Sep	Oct	Nov	Dec	Total
0	5,000	15,000	20,000	20.000	20,000	120,000
1,000	3,000	2,000	5,000	4,000	6,000	50,000
4,000	2,000	1,000	3,000	2,000	3,000	35,000
25,000	15,000	2,000	17,000	24,000	26,000	225,000
30,000	25,000	20,000	35,000	40,000	45,000	430,000
20,416	20,416	20,416	20,416	20,416	20,416	244,992
4,491	4,491	4,491	4,491	4,491	4,491	53,892
0	0	0	0	0	0	15,000
0	3,000	4,500	0	0	0	7,500
2,000	2,000	2,000	2,000	2,000	0	22,000
500	500	500	500	500	0	5,500
500	500	1,000	1,000	1,000	0	13,500
1,093	1,093	1,093	1,093	593	93	12,616
11,000	13,000	8,000	5,500	1,000	0	55,000
40,000	45,000	42,000	35,000	30,000	25,000	430,000
12,000	(8,000)	(30,000)	(30,000)	(20,000)	-0-	

■ *How are we supposed to set up an annual cash flow budget?*

First, create headers for each item in your budget. Items such as grants, donations, and earned revenue are listed under the general heading of income. If you have an item in your chart of accounts for fringe benefits, use the same term in your cash flow budget.

Avoid using code or account numbers. Most of the people reading this budget will be baffled by them. If you wish to refer to these codes, simply run off a separate budget sheet for staff that includes them, but leave them off for the rest of us.

Number each item in your budget. These numbers will serve as footnotes containing additional information. For example, using the sample cash flow budget, this information would be shown on a page of footnotes, each corresponding to the number of the item on the budget page:

5. Salaries	$245,000
6. Fringe Benefits	$53,900
7. Part-time Wages	$15,000
8. Consulting Fees	$7,500

On the footnote page, provide the reader with the formula, the specifics, and all the relevant information that is needed to understand your budget. This is particularly helpful to those unaccustomed to reading financial statements or reports. This includes a majority of board members and most other reasonably sane members of society.

Translate your cash flow summary sheet into a chart or a graphic format. The shape of your income and expense curves will be more meaningful than lists of numbers to some readers, particularly the 60% to 70% who are likely to be visual learners. The graphic format demonstrates the relationship between income and expense in relation to time.

When developing the income side of your cash flow budget, clearly differentiate anticipated funds using three categories:

Secured funds: money already in the bank, part of a multi-year grant, or backed by contracts or collectible pledges.

Highly probable funds: funds that you can historically count on coming in from major contributors or agencies with which you have long-standing relationships.

Speculative funds: proposals pending, or projected income from a campaign that is planned but not yet implemented.

You can use your footnotes to indicate in which category the estimated funds currently fall. This will enable you to gauge the relative strength of your position.

For example, you might note that 30% of your funds are secured, 50% are highly probable, and 20% are speculative. In this case, you are in a relatively strong financial position. On the other hand, if 75% of your budget is speculative, you may want to be very careful about your expenses during the first two quarters of the year.

Easy-to-Understand Budgeting

■ *Why are footnotes needed to understand this budget? Shouldn't the numbers be self-explanatory?*

Financial reporting in the nonprofit sector is not as standardized as you might think. Exotic and creative terms, as well as interesting and sometimes arcane concepts, seep into even the most conventional fiscal documents. Since most board members have better things to do with their lives than learning a new financial language, it makes sense to use footnotes to spell out exactly what each term means.

For example, in our cash flow budget the following footnotes would be included:

1. Grants. The total amount of grants anticipated this year is $120,000. Of this amount, $50,000 has been secured as part of a multi-year grant from the state. $40,000 is highly probable, since we are asking for funds from agencies and foundations that have demonstrated a strong interest in our activities or have shown an historical interest in funding our programs. This leaves $30,000 that is speculative. By the end of the first quarter, we will have proposals pending to five new foundations and six corporations.

2. Donations received as part of our annual auction and year-round volunteer telemarketing campaign. Last year the auction raised $15,000. It should do that well or better this year. Consequently, we are forecasting that the auction will raise $20,000 this year.

The telemarketing program has generated a fairly consistent $2,000 to $3,000 per month, with the exception of a sharp dip in midsummer and with some higher numbers in October, November, and December. We believe that at least $20,000 can be considered highly probable from this source.

The balance of the $50,000 scheduled for donations is speculative. We believe that the auction and the telemarketing will deliver these funds above and beyond our estimates, but we cannot be certain of achieving this goal. To increase our chances of reaching our mark, we will be using more part-time helpers for the auction this year and will be placing emphasis on obtaining more items and greater attendance at the event.

3. Gifts. We differentiate between donations, which are contributions made as a result of direct solicitation through telemarketing or as income from our annual auction, and gifts, which are contributions that come directly from friends or associates of the organization. Most of our gifts come from planned giving. Although it is difficult for us to estimate the arrival of funds, our experience indicates that roughly $25,000 per year will come from this source. The balance of the $35,000 allocated to this category, while speculative, should come from a program that was initiated last year. By asking board and staff to secure matching funds for their gifts to the organization, we are confident that we will obtain $10,000.

4. Earned Revenue from the services we provide is a constant source of cash. The significant swings in income, from a high in June to a low in September, are

characteristic of our annual service cycle. Our line of credit from First National Bank enables us to stabilize our operational cash flow and meet our monthly obligations.

5. Salaries are provided for members of the staff. The breakdown by position:

Executive Director	$ 40,000
Assistant Director	$ 35,000
Clinical Supervisor	$ 25,000
Development Director	$ 20,000
Two Clerical Staff at $16,500 each	$ 33,000
Four Field Workers at $23,000 each	$ 92,000
Total Salaries	*$245,000*

6. Fringe Benefits are calculated at 22% of base salary. They include our health plan through Group Health, the pension plan we offer through Metropolitan Life, and the portion of Social Security paid by the organization.

($245,000 x .22 = $53,900) Divided by 12, this amount equals $4,491 per month.

7. Part-time Wages are projected for assistance with our annual auction. This year, it is estimated that we will require up to 2,500 hours of effort. Based on an average cost of $6 per hour for this help, we have budgeted $15,000 to be spent between January and March.

(2,500 x $6 = $15,000) $ 15,000

8. Consulting Fees. Our strategic plan is being developed by a planning and consulting firm that will work for a total of 50 hours on this project.

(50 hours x $150 per hour = $7,500) $ 7,500

9. Rent has been budgeted for 11 months this year. In December we will be moving into a rent-free space that has been made available to us by a local real estate developer who has room in one of his buildings. We can use the space for two years. At the end of this period, we will be asked to pay market-value rent.

10. Utilities have also been budgeted on an 11-month basis. The new space will be provided to us without utility charges for the first year.

11. Legal and Accounting services are higher than usual this year. We will be audited for the first time and will incur higher fees during the first three months of the year. Our normal legal retainer is $500 per month, and this year an additional $500 has been included for accounting services during the months of September, October, and November. No payments are scheduled for December.

12. Supplies have always been an important expense for us. The supplies are vital to the service we perform. With the exception of January, which is usually a costly month, our normal monthly cost is $1,093. During May, November, and December, our inventory of supplies is always allowed to decline.

13. Printing continues to be a major expense item because our publications are in considerable demand. Printing schedules are timed to coincide with our major program activities.

When this information is included in the form of footnotes, board and staff members can see the numbers and the logic that informs them. This lends considerable clarity to

23

budgets, and it also helps to illuminate shaky assumptions that may need to be challenged before things get out of hand.

■ *What is a summary cash flow budget, and why is it important?*

The following is a summary annual cash flow budget. The form has been pared down to the absolute minimum, so that only monthly dates, income, expense totals, and the running total are presented. People who want more detail can easily reference the annual cash flow budget or the monthly cash flow projections that have been used to create this summary.

	Jan	Feb	Mar	Apr	May	Jun	Jul	Aug	Sep	Oct	Nov	Dec	Total
income	30	32	38	45	50	40	30	25	20	35	40	45	430
expenses	41	40	35	32	30	35	40	45	42	35	30	25	430
running total	(11)	(19)	(16)	(3)	17	22	12	(8)	(30)	(30)	(20)	0	0

(All figures are in thousands)

This minimal budget, showing only income, expense, and a running total, can be used to quickly convey information to the reader. It also provides the basic information necessary to generate charts or graphs to help some readers better understand the dynamic quality of your annual cash flow.

Financial Stability

■ *How can we avoid the management-by-crisis mode of operating?*

Perhaps the single least understood method for stabilizing the financial situation of a nonprofit organization is the use of credit or borrowing. The management-by-crisis mode of operating ultimately comes down to not having the funds you need to pay the bills you have, when you need to. Businesses across America routinely borrow when they need funds and pay back when they don't. Yet for many in the nonprofit sector, borrowing carries with it an undeserved stigma and is viewed as a sign of poor management.

An underlying premise of this guide is that if you have the ability to establish a line of credit with your bank, then you have the essential financial management skills to ensure your fiscal stability. Borrowing is our metaphor for all the effort necessary to forge a working understanding of credit with the board of directors, the formation of a community-based credit holders group, the preparation of clear and understandable financial statements, and the ability to project cash flow. It also involves selling the organization and its fiscal position to a sometimes reluctant banker.

The way that businesses stabilize their cash flow is by using a line of credit. In other words, they borrow when they need cash and repay when they don't. The same applies to non-profit organizations that wish to stabilize their cash flow. A line of credit equal to 10% to 20% of the annual operating budget should be a goal for every nonprofit organization.

Credit Holders

We don't have collateral to secure a line of credit. How can the bank lend to us without collateral?

Bankers cannot lend to you without collateral. Nonprofit organizations trying to establish a line of credit discover that banks require two guarantees of repayment (cash from operations and assets to act as collateral) before they are willing to lend. Typically, the primary source of repayment is from a nonprofit's earnings or from grants and gifts. The problem for many institutions is that they lack assets that bankers can easily convert into collateral.

In this regard, you are not alone. Few of the "things" that nonprofits have are easily converted to cash. Trying to sell them could be hard for you and expensive for your banker. But you may have access to a form of collateral that you have not considered, one that is within your reach and that bankers worldwide are happy to accept without reservation. That collateral is generated by credit holders and is called cash that is pledged to enable you to borrow against it.

We suggest that organizations set up a "credit holders" group. Credit holders are friends, associates, supporters, family members, staff members, or constituents who help by placing a small amount of their savings into a special interest-bearing account at the bank.

Usually, these funds are used to purchase a certificate of deposit. The money in the account is pledged to the bank as collateral for borrowing by the nonprofit agency. Each credit holder receives interest based on the amount they deposit. Interest is paid out annually or rolled over into the account.

Before soliciting funds from credit holders, the board needs to establish a policy stating that this line of credit will be used only to cover fluctuations in cash flow, not to address long-term debt or to serve as venture capital. This means that before the credit holders' funds are pledged as collateral, the institution will already have demonstrated the capacity to repay the loan from operations.

Given this policy, the degree of risk to credit holders is kept very low.

You and your board should talk face-to-face with others about becoming credit holders. After you explain the organization's borrowing policy and outline any possible risks, each person is asked to place a modest sum into a collateral account. This is done with the understanding that, even though risks are low, losses do happen.

The incentive to support the organization is hardly financial. After all, people can earn interest with complete security in a bank savings account. Rather, by placing their funds in a special account with your organization, each person will be helping you to operate effectively and further the social purpose of your organization, at no cost to themselves.

Obviously, it makes sense to check with a lawyer to ensure that no state laws are violated. To date, credit holders groups have encountered no legal obstacles, but it is wise to check out your local situation before proceeding.

Interestingly enough, credit holders often become the bottom tier of the organization's fund-raising pyramid. After people have participated for a while, they often feel a sense of partnership with the organization. Credit holders, in addition to helping provide credit, start to make donations.

This method offers several advantages to the parties involved:

> The organization gains a stable resource that can be used to mobilize credit whenever it can be shown that other sources of revenue will repay the loan.

> Interest charged by banks for lines of credit secured by credit holders' certificates of deposit are typically 1 or 2 points above the interest paid by the bank for the certificates of deposit. This is usually 2 or 3 points below prime rate.

> Borrowing at a lower rate enables nonprofits to pay their vendors quickly, gain discounts, and avoid late payment penalties. The discounts can often cover the interest costs of borrowing.

> Credit holders gain an opportunity to provide invaluable assistance to a nonprofit organization they believe in, without cost to themselves.

> The bank gains collateral it needs to justify a loan to the organization.

Meeting Your Banker

■ *We have a credit holders group; now how do we approach the bank for a line of credit?*

You have prepared your annual cash flow budget, complete with footnotes. Furthermore, you have convinced your trustees and staff to institute a credit holders group. They agreed to make their own funds available and to get others to participate. You are able to quickly determine the amount of money you will need to borrow by looking at your cash flow budget and noting the largest negative monthly total. Or you may anticipate future borrowing needs by simply assuming that your line of credit should equal at least 10% to 20% of your annual operating budget. What do you do next?

■ *How do we prepare before we meet the bank?*

Only death, pestilence, and taxes inspire as much fear as visits to the bank to ask for money. So you will want to sit down with your financial committee chairperson or your board treasurer and review the financial statements you will need to present to the bank. You should take some time to quickly sketch out a strategic plan, a marketing plan, and a brief description of how you will pay for expenses in the coming year. These need not be exhaustive planning efforts, but they do need to accurately reflect your best thinking at the time.

This preliminary work will pay a handsome dividend to you in the future, not only at the bank but in your fundraising efforts, in your day-to-day operations, and in your relationship with staff and trustees. Organizations using a consultant to assist them in preparing these materials have found that costs for the service are normally quite modest.

When all your homework is done, you will have:

> An idea of how much you want to borrow.

> Financial reports and plans to show how you expect to repay the loan.

> A credit holders group.

> Enough funds to match the amount you wish to borrow, plus a little extra to allow individual credit holders to withdraw in an emergency.

In sum, you are ready to go shopping for a bank that will recognize your worth, offer you good terms, and work with you to meet your credit needs in the coming years.

■ *How do we make a date with the bank?*

Start by calling the bank where you normally do business. Ask to set up an appointment with a loan officer and indicate that you are interested in establishing a line of credit. Don't be surprised if you encounter some resistance from the people at the bank. A remarkable number of bankers have no idea that nonprofits qualify for, and can use, a line of credit. They may think that you are looking for a donation or an unsecured loan.

Just be patient and remember that, while they are in the business of renting money, bankers are sometimes naive when it comes to dealing with nonprofit organizations.

30

■ *What's going to happen at our first meeting?*

This depends on the banker, but don't count on too much. The first meeting with your banker should be viewed as a "get acquainted" session, with a specific request for a line of credit and some paperwork for the banker to review. Your banker will be most concerned with whether the loan can be repaid. He or she will want to evaluate your current financial position, your prospects for the future, and the quality of your management.

When meeting with your banker, the following items are helpful to take along:

1. Three years of audited financial statements. These statements, if you have them, provide information on your fiscal solvency.

2. Organizational plans:

Strategic plan: demonstrating the fit between the organization and its environment.

Assets management plan: plans for managing your cash and other assets. You might include your policy toward vendors here.

Marketing plan: a plan for identifying and reaching your audience or clients.

Financing plan: a fund-raising plan, a plan for generating earnings, and a plan for borrowing. If you have formed a group of credit holders to provide collateral, this is a logical place to describe how it will work.

3. Prior years' budgets. Budgets are forecast statements of operations and financial position.

4. Current annual cash flow budget. At the very least, this should include forecasted statements of operations, financial position, and projected cash flow. The projected cash flow statement is especially useful in identifying whether there will be sufficient cash to repay the loan.

Be prepared to answer questions, but remember that there is no need to be defensive. The bank is renting money, and they have a right to ensure that you understand your responsibility to repay it. However, since you are offering cash as a source of collateral, through your credit holders' account, the banker should be very accommodating. He or she should be prepared to offer excellent terms on your line of credit, since the bank has little or no risk associated with your loan.

If you feel uncomfortable during this transaction, tell the banker. You are offering the bank an opportunity to earn income on your borrowing with virtually no risk, and you should be very well treated. If the experience or the terms do not suit you, consider going to another bank. Businesses shop around for banks all the time. While the products offered are pretty much the same, the treatment and service provided vary greatly.

■ *Is it more wonderful the second time around?*

There may be many details to discuss at your second meeting with the banker. For example:

How will the credit holders' account be established and administered? Will the bank compute the individual interest owed to the participants, or will the account be set up in the name of the organization and the responsibility for figuring the interest left with your treasurer?

What is a fair rate of interest to be charged for borrowing on the line of credit, particularly given the low level of risk to the bank? Will the banker waive the 1% or 2% fee normally charged for maintaining a line of credit?

Will the line of credit grow with your annual budget after the initial year?

How will you have access to the funds in the bank? Will you need to visit the bank to transfer money into your account, or can this be handled over the phone?

Does the bank want to have receivables assigned to them for collection? For example, should you send a grant from a government agency directly to the bank as repayment for borrowing?

■ *Can we just sign up for financial stability?*

Only after these thorny questions have been resolved will you be prepared to sign up for a revolving line of credit with your bank. After securing your line of credit, make sure you leave a little time to celebrate. You will have achieved a significant milestone in the financial management of your organization. From now on you will have

funds available to you to meet your cash flow requirements. Late checks from foundations or government
agencies will no longer result in anxious moments as
payday approaches. You will have access to a resource
that is renewable for as long as you fulfill your end of
the bargain.

Deficits

■ *We have suddenly discovered a huge deficit. Before we burn the executive director at the stake and explore Chapter 11 proceedings, is there anything we can do?*

Before running for the firewood and matches, you need to correct or take care of the problem created by the deficit, and you need to do it in a way that is least harmful to the organization. Shakespeare was wrong: Hell hath no fury like a board member who feels betrayed by suddenly learning of a deficit. Be gentle; find the least painful way to solve the immediate financial problem, and at the same time begin to put into place or refine systems designed to prevent a similar situation in the future.

■ *Are you suggesting more debt?*

Yes, but first try to look calmly at the deficit. How big is it? Could it be brought under control if converted to term debt and paid off, just like a mortgage over five or six years, with reasonable interest payments?

Term debt is borrowing that is stretched out over a fixed period, say 60 months (five years), with regular payments that are divided up into interest and principal. Your line of credit needs to be completely paid back at least once during the year. This ensures that it will be used only for timing and not for debt or venture.

The advantage of term debt when you have a deficit is that it allows you to slowly and deliberately pay off your obligation. The obvious disadvantage is that your deficit may be an indication of underlying operational problems that need to be resolved before you can make money in the future. You and the staff need to make this determination. You also need to put together enough collateral to persuade the bank that it will be protected if it rents you money over a longer period of time.

This may be the time for the board members to establish an internal credit holders group, as discussed earlier. By offering the banker a liquid asset to collaterize the term loan, you are shouldering the responsibility for curing the problems that caused the deficit.

After analyzing their situation, many of our clients have repositioned their deficit with their banks. Having a secured line of credit and some history with the bank saves time and energy, and ultimately makes the probability of obtaining a term loan much higher.

■ *For heaven's sake, what do we do then?*

Once the bleeding has been stopped, it makes sense to carefully explore the reasons that the deficit occurred. Has it been growing for some time? Given the vagaries of fiscal reporting in many nonprofit organizations, things could have been cooking for a while before anyone smelled it burning. Look first to the financial reports. Do they reveal a problem? Often a trend has been in the making, but it was not candidly disclosed to the board or was ignored in an act of massive denial by members and staff.

Does the deficit follow Linzer's Law of Littles? To wit, did lots of little items, such as small losses of income or small extra expenditures, come together to form a sizable deficit? If things are subtle, and with two bottom lines to address, small items can gradually create big deficits. One answer is to refocus in just the opposite way, to look carefully at the big picture.

Return to an examination of cash flow on a monthly basis, looking at total income, total expense, and the running or cumulative total. These numbers can often help to point out when things started going south.

Does the deficit result from the postpartum letdown that accompanies or follows an ambitious capital campaign? Over the years, we can't tell you how many times we've heard board presidents or executive directors moan about the deficit in operating income that resulted just after a capital campaign. You are trying to feed two mouths when you raise capital, and without tremendous effort the operational budget, the daily grind that no ones loves to contribute to, will come back to haunt you. Beware the capital campaign: it can be a real deficit maker!

■ *Did someone really make off with the funds?*

Is there an address in Bora Bora where you could send a note inquiring about the loss? This sounds funny, but it isn't. People who work for nonprofit organizations may be devoting themselves to the public good and social purpose, but some of us still harbor felonious tendencies. The commercial world is not the only place where people embezzle funds. Good controls and common-sense ideas, such as separating the function of opening the mail from the task of posting checks to the account, can help. Trust me; even though the accountants are still one step behind the crooks, they still have excellent advice to give in this department. The accountants, that is.

■ *If our organization gets into financial trouble, shouldn't some members of the board lend the organization money out of their own pockets?*

When my father wanted to buy a car, he went to his father and asked for a loan. My grandfather was a businessman who understood the value of a dollar, but he did something that in retrospect makes good sense. He took my father down to the bank and made arrangements for him to borrow money from the banker to buy a new car. As he explained to my father, "I value our relationship much too much to allow issues involving money to affect it. Therefore, even though it will cost a few dollars in interest, you make the payments to the bank."

For much the same reason, we suggest that board members not lend funds to their organization. Your relationship with the institution is much too important to be cluttered up with personal issues regarding its financial situation. In our experience, board members who lend money often propose the most draconian measures of austerity when times are tough for the institution, sometimes to their own and the organization's detriment. As board members you can donate money, help with fundraising, pledge funds as collateral, and get others to join you. All these are positive acts. Let the bankers lend, stop worrying about the marginal cost of interest, and get on with the business of the organization with a clear conscience.

Capital Accumulation

■ *Americans have always accumulated capital. Why shouldn't
we try to accumulate capital?*

Capital accumulation is the name of the game in American
life, so it is not at all surprising that it should be the
dominant focus in development work for nonprofit organizations. Capital campaigns that are designed to raise
funds for cash reserves, endowments, new buildings,
renovations, or equipment purchases are forms of capital
accumulation. Yet capital accumulation within the nonprofit sector confronts us with some interesting paradoxes
and problems.

Despite its popularity, there is growing evidence that
capital accumulation by nonprofit organizations is harmful for individual institutions, the constituencies they
serve, the local community, and ultimately the entire
nonprofit sector. While larger institutions worry about
having adequate resources, the ways in which they handle
the money they acquire is often very inefficient. As a
consequence, these institutions offer fewer services relative to their resource base. At the same time, the larger
institutions' need for more money, in the face of an inelastic pool of funds, squeezes the smaller institutions, who
provide a higher level of service in relation to their overall
budget.

A fast review of contemporary economic theory helps to
explain why this happens. Let me clarify an important
point by noting that the economic environment in which
we operate has changed dramatically in the past few

years. Not only are we moving from a nationally based manufacturing economy to a globally based information economy, but the way in which economists see the workings of the economy has profoundly changed.

The economics that we grew up with was called the dismal science, in large measure because economists cling to the concept of diminishing returns as the fundamental element in economic affairs. Keep planting the field year after year and production of crops will decline. Find a niche in the market, manufacture products to fill the niche and compete like crazy until equilibrium occurs, and then move on to the next product. This is the economic version of the second law of thermodynamics. "All things run down." "All things diminish." "Entropy reigns supreme."

■ *Has anyone other than Bill Gates ever heard of increasing returns?*

For some of us, the concept of diminishing returns has been inherently unsatisfying for some time. All around us, we see ample evidence of a phenomenon called "increasing returns," and we need to factor it into the equation.

The law of increasing returns says simply, in the words of the King James translation of the Bible, "To them that hath shall be given." In other words, them that has, gets, or the rich get richer. This means that having something, be it market share or capital, tends to promote behavior in the system that brings increasing returns to the owner. Microsoft is an excellent example. Bill Gates appears to have intuitively understood that if he could get a large enough share of the market within the field of information

40

technology, he would ultimately dominate the field, even with less superior technology.

Do you remember Beta versus VCR? Or how about the 24-hour clock face versus the 12-hour face on your watch, or even the Qwerty keyboard on your computer? All of these are examples of increasing returns, in which market share won over superior technology.

Increasing returns also seem to operate in the nonprofit sector. The remarkable growth of public education and the modern health care system are two examples of how, as you supply facilities and practitioners in the service sector, you can increase the demand for services until eventually the demand exceeds the supply, at least in the short term. Instead of satisfying the market, your services stimulate behavior that causes the market to expand, reinforcing its own growth.

■ *Anxiety about the future is connected to all this, isn't it?*

In a world in which there were only diminishing returns, capital accumulation strategies might make sense. When we interview the leaders of large nonprofit organizations, they frequently tell us that their concerns for future funding are tied to diminishing resources. So it appears that there is a powerful perception among these institutional leaders that their organizations are endangered. Yet the theory of increasing returns tells us that the rich get richer; indeed, by and large, the leaders interviewed will confess that their organizations continue to grow each year.

To understand the roots of this anxiety about the future, it is helpful to note that giving across the United States is remarkably constant. The pool of grants and donated dollars is relatively stable. Studies over the past 30 years have shown that the amount of money contributed each year has grown at slightly less than the rate of inflation. This is true if you consider income from all sources, including individuals, corporations, and government agencies. Studies conducted annually by national fund-raising associations show that contributions each year tend to be very stable. This was true for 1995 and 1996, when a strong market, low inflation, and generational transfers of funds created a banner year for individual contributions. But the overall pool of resources remained constant because corporations and government gave less.

While the pool of resources is relatively stable, the demand for resources continues to grow, if for no other reason than the fact that inflation in the nonprofit sector is one to two full percentage points higher than in the rest of the economy. Plus, new nonprofits are forming all the time. In fact, the number has tripled since l970, and existing organizations continue to require fresh infusions of capital.

So it is small wonder that leaders of midsized and larger institutions worry about all those small fries, despite having increasing returns on their side.

If there are increasing demands, and if the level of giving is stable nationwide, then who is going to get the lion's share of the donations?

If we believe in diminishing returns, then the larger institutions have good reason to be concerned, since every dollar that they do not capture will go to another institution. However, if we believe in increasing returns, then the lion is going to get the lion's share, regardless of the amount of demand.

Since the pie is relatively fixed, the issue becomes one of either finding ways to make each piece go further or having an increasing number of agencies and organizations go hungry.

You can see the problem. The biggest institutions are going to grow. After all, the theory of increasing returns states, "Them that has, gets." This is true even when these large institutions utilize their capital inefficiently with endowments, cash reserves, buildings, and equipment. They will constantly need more funds just to meet their inflation-driven operating expenses. What this means within a community is that the smaller organizations will be starved for resources as their larger peers continue to consume more and more capital. If the pool of resources does not increase, the competition will increase and further disadvantage the smaller institutions.

Imagine a family about to sit down for dinner. The food is on the table, but even before grace can be said, the hulking teenage son grabs the steak and the chocolate cake, takes them to his room, and returns to the table empty-handed. He says, "Gosh, I'm hungry." And to the amazement of everyone else, he bows his head.

When asked what he intends to do with the food in his room, he answers, "Eat it later." Everyone whines, "That meal was supposed to be for all of us." "Sorry, but I'm a growing boy," he says, continuing to sit at the table with the rest of the family, ready to eat the few dishes that are left.

Not a pretty picture, but not that different in our mind from what happens when institutions accumulate large amounts of money, invest it for long-term gain, and still have to compete for resources to satisfy their current requirements.

■ *What are endowments?*

An endowment is an irrevocable trust, a legally binding agreement in which funds are given to an institution with the understanding, in most cases, that the principal is to be maintained in perpetuity. Only a portion of the income earned can be spent. In addition, institutions accumulate capital through cash reserves and funds restricted by the board, often called quasi-endowments.

As we mentioned earlier, nonprofits have two bottom lines: the immediate pursuit of their social purpose, and long-term fiscal solvency. An endowment is a financial device that defers the immediate gratification of current operational needs for the imagined long-term benefits of fiscal solvency. The financial strategies that organizations adopt, including endowments or lines of credit, are simply ways to confront the need to deal with financial solvency over time.

■ *What's the other side of the coin with regard to endowments?*

Endowments are very popular. Almost everyone in the nonprofit sector wants an endowment, a cash reserve, or a capital campaign to construct a new building or to renovate their existing space. In other words, board and staff members are pursuing strategies to accumulate capital for their institutions. Yet endowments are rarely an efficient way to address issues of fiscal solvency. They are costly to raise and contribute relatively little to the annual budget.

Many privately held endowments in this nation spend less than 5% each year to fulfill the mission of their organization. In this day of 3% inflation, if we add the additional two points that inflation in the nonprofit world is assumed to have, you can see that even at 5%, the endowments barely contribute enough to match current inflation, to say nothing of operational needs. If inflation increases again, despite the efforts of the Federal Reserve Bank, the performance of endowments will decline even further.

■ *But won't we be able to borrow against our endowment in times of need?*

An irrevocable trust (an endowment) is frequently useless in the face of fiscal crisis. You can't touch the principal, and neither can your banker, so you can't borrow by using it for collateral. For this reason, it is technically possible for a nonprofit organization to be, at the same moment, fiscally solvent (at least on paper) and bankrupt. Since you can spend only a portion of the interest, you may have an impressive endowment, but you won't be able to mobilize the money to pay your bills. In this case, interestingly enough, the courts will not liquidate the endowment. Instead, they will transfer the principal to another nonprofit institution, leaving trustees still on the hook to creditors.

■ *If the pool is constant, isn't that all the more reason to get in there and get our share for an endowment?*

Inefficient use of funds harms everyone: the institution, the community, and the clients and constituents of non-profit organizations. When excess funds are devoted to endowments, there are fewer funds available to meet the organization's operational needs. This means that clients and constituents receive less service or have to pay a higher price.

In addition to endowments, restricted funds, and cash reserves, capital can also be accumulated in the form of buildings and equipment. The ownership of property and tangible goods is very much part of the tradition of the nonprofit sector. This use of capital is also inefficient.

■ *Hey, hold on there. If endowments are so bad, why does Harvard have a $14 billion endowment fund?*

It's true that Harvard has a $14 billion endowment fund; it also just spent $15 million to build an athletic facility without showers. And the fact that many institutions have endowments doesn't mean they are bad; rather, endowments are a highly inefficient use of capital, and that is bad.

■ *How do we figure the cost of an endowment versus the benefits to us?*

As an example of the fiscal efficiency of endowments (irrevocable trusts), this quick formula helps to assess the amount of funds available each year to fulfill the mission of the institution.

46

Using $100,000 as a basis, apply the following assumptions:

Goal of the endowment campaign	$100,000
Estimated cost of campaign (33 cents per dollar)[1]	33,000
Annual yield[2] (8%)	8,000
Annual amount returned to principal[3] (3%)	(3,000)
Correction for inflation in the nonprofit sector[4] (2%)	(2,000)
Balance of annual spendable income from the fund	3,000

To recapture the sum spent to raise the endowment, in this case $33,000, the institution will not receive a penny of appreciation for 11 years.

[1] The Association of Health Care Philanthropies has studied this question for years. Their figures indicate that nonprofit organizations with full-charge development offices (meaning those that spend more than $180,000 per year on this function) will spend a median of 33 to 77 cents to raise a dollar. The major variable is time. Development efforts that have been in existence for more than 10 years spent a median of 33 cents, while those less than two years old spent a median of 77 cents.

[2] Funds invested 60% in stocks, 30% in government securities, and 10% in cash are considered prudent. Over the last 20 years, the return on investment has averaged 5%.

[3] To sustain the buying power of the fund, an amount equal to an estimated five-year rolling average of inflation needs to be returned to the fund. The current estimate for inflation is 3%.

[4] As Baumol and Bowen noted in their classic study of the economics of the performng arts, inflation in the nonprofit sector is one to two full percentage points higher than in the commercial economy. They cited three reasons: (1) the nonprofit sector is more labor intensive, (2) the smaller economy of scale for nonprofits produces higher costs for these organizations, and (3) there is no incentive to invest in advanced technologies within the tax-exempt environment; consequently, nonprofits have lower levels of technology with which to compete.

■ *We don't spend that much to raise a dollar, do we?*

Clearly, one of the key assumptions included in this formula is the cost of raising a dollar. We know that the actual cost of raising a dollar may vary. It is often difficult to make a judgment about this because there is so much pressure in the philanthropic community to report low costs for development efforts. Institutions either under-report their costs or face losing support from donors.

If you pencil in the actual costs of raising money, you will see one reason that endowments are an inefficient fiscal device. If the total cost of raising endowments or cash reserves is included, it may well take years before institutions realize a penny of positive appreciation.

In other words, the history of endowments in this nation shows a poor track record of meeting the needs they are supposed to address. The spending policy of most trusts simply maintains the status quo.

■ *A wealthy donor has offered us a challenge. He will give us a million bucks to establish an endowment if we match every one of his dollars with three that we raise in the community. Shouldn't we jump at the offer?*

You might want to look before you leap. Calculate the cost of raising the three-to-one match, particularly on top of your normal operating expenses, then take a hard look at the spending policy you will need to adopt to sustain the buying power of the funds, and finally figure out how long it will take before you see a penny of appreciation once you have recaptured your original investment.

Then, before accepting the donor's idea of what is good for you, ask yourself whether the donor needs to examine those numbers with you to see the consequences of his or her request. Show the donor the relatively high cost of raising funds, the relatively low return available from endowments, and the long-term inflexibility of irrevocable trusts as instruments to prevent financial disasters.

In our experience, a donor will often, when confronted with the facts, change his or her mind and offer you a much better deal. For example, the donor may decide to donate the money to be used for operations, with the understanding that you will match it three to one for the same purpose.

■ *If we have an endowment, it's easy to approach donors. How can we convince them to support us if we don't have an endowment? Who would give to us? Won't we lose money and community support?*

While it may be true that everyone wants an endowment, a cash reserve, a new building, or a renovation, the competition for capital gifts is starting to have a profound and negative impact on the nonprofit sector. A recent study of capital giving to San Francisco nonprofits from 1988 to 1992, commissioned by the Walter and Elise Haas Fund, points out that the demand for capital giving may soon outstrip the supply. They reached five main conclusions:

> Capital giving is growing significantly faster than overall giving.

The aspirations of San Francisco nonprofit organizations for capital donations would appear to outstrip future donors' giving capacity.

Many campaigns will likely fail or fall significantly below their targets; alternatively, if they succeed, there will be a significant shift in charitable giving that may negatively affect many organizations.

Foundations played a major role in the growth of capital giving. This increase was made possible by unusual growth in assets in the period under study.

Capital campaigns have the effect of redistributing total charitable giving, reducing income to some recipients and increasing income to others.

■ *If endowments aren't such a great deal, why does everyone else have or want one?*

The concept of an endowment is a bit like chocolate cake: wonderful to dream of and delicious to eat, but not always good for you. Just because almost everyone likes chocolate doesn't remove some of the side effects of eating too much chocolate.

There are a number of negative consequences of creating, developing, and sustaining endowments. Endowments are costly to raise on an ongoing basis, require substantial care and feeding, and produce relatively tiny amounts of spendable income, even after your investment has been recaptured. Endowments take 95% of the money that could be used for fulfilling your mission and place it in investments that are forever outside your reach. In addition, since the pools of resources they are drawn from are relatively inelastic, they deprive others of needed resources. They do not satisfy your needs for operational funds. They can result in diminution of services. Clients and constituents may suffer.

■ *Still, won't it be nice to have a large endowment so we won't have to fund-raise each year?*

Since the operational needs of the organization continue year after year, and since endowments contribute, relatively speaking, very little to most institutions' operational budgets, the need to raise funds is not diminished by endowments. Although many donors are asked to contribute on the promise that this is a once-in-a-lifetime gift or a one-time special effort, the reality for most institutions is that they will be back at the door the next year, still needing operational funds to fulfill their mission.

■ *Having a rainy-day fund is very desirable, isn't it? And, after all, why would we want to pay all that interest to banks?*

Cash reserves or rainy-day funds always seem like a wonderful idea. The thought of having a little extra cash tucked away for a cold and blustery budgetary lapse seems so cozy and warm. And, frankly, it is nice to have some cushion in reserve. But does it make sense to have cash lying around that is expensive to raise and will produce very low rates of return given its liquidity? Credit serves exactly the same function for a nonprofit organization, and it allows you to place your hard-earned dollars into service of your mission anytime you have a need. Borrowed dollars are current, and they are inexpensive to mobilize. If you have a credit holders group in place to secure your line of credit, consider letting your banker be your rainy-day funder, and use the other capital more efficiently.

■ *Doesn't it make real sense to adopt businesslike practices and defer paying our vendors for as long as possible?*

The notion of withholding payment of bills makes little sense in the nonprofit sector. Goodwill is much more important for most nonprofits. They lack the economic muscle to stiff vendors and still keep them coming back for more. The loss of discounts and the assessment of penalties for late payment are rarely worth the advantages of withholding payment. Organizations that create a credit holders group and establish a secured line of credit will find that they can pay on time, reduce or eliminate penalties, gain valuable discounts, and secure far better terms with vendors. Some organizations that use a secured line of credit to pay quickly even ask their vendors to consider writing a contribution check instead of a discount. Since the vendor that you pay promptly, on delivery, loves you more than your dog, some vendors are willing to make out tax-deductible contribution checks for up to 10% of the bill.

■ *Our development officers tell us that planned giving is the way to go. Is this correct?*

Aspects of planned giving are attractive for those nonprofit organizations that have the capacity to track donors over long periods and the discretionary dollars to provide educa-

52

tional programs for donors. Usually attorneys are brought into the process by planned giving consultants, and the costs for services can mount rapidly. Bequests are almost always welcome as gifts, and some such donations, unlike gift horses, do not require extensive dental examinations before being accepted. However, some of the more exotic charitable remainder trusts and annuities need to be carefully scrutinized, since the benefits to the donors in the present and over time may cancel out the value of the gift for the recipients.

■ *Going further into debt seems like a lousy idea. Shouldn't we embark on a massive fund-raising effort to cure the accumulated deficit that has recently come to light?*

Doesn't it make sense to first ask why you waited until accumulating a deficit to embark on a "massive" fund-raising campaign? If your fiscal reporting did not warn you that a problem was accumulating, that should be a real concern and needs to be addressed. Then, after examining the problem, it may make much more sense to translate the deficit into term debt and develop a strategy to pay it off over time. Some organizations have found that crisis-driven fund-raising, the kamikaze approach to development efforts, will work in the short run, but that it is hard to sustain the momentum over the long haul. Fund-raising needs to be viewed as a disciplined, ongoing activity, not the end product of poor financial monitoring and inadequate planning for the credit needs of the organization.

Ownership by a Nonprofit

■ *Owning our own building will solve a lot of problems, won't it?*

It is currently fashionable for nonprofit organizations to purchase buildings or renovate buildings they own. Yet few institutions take the time to carefully analyze the costs and benefits of long-term leasing versus ownership before they start their capital campaigns.

Mariners say that the two happiest days of their lives are the day they buy a boat and the day they sell it. Perhaps the reason for this is that—unlike homes, which appear to have substantial value as equity—boats tend to decline in value, like cars, and still cost the owners more and more to maintain each year.

In general, real estate ownership is economically justified because of tax advantages resulting from the deductibility of expenses—such as mortgage interest, depreciation, property taxes, and maintenance—from income. But nonprofits are tax-exempt, so they obtain no tax advantages from ownership.

For nonprofits, it is the use of a space—the control over the space and the predictable annual cost of the space—that is important. So while the maintenance and expense of owning a building are similar to those involved in owning a boat, ownership of a building, as we understand it in the commercial world, is not even an option for nonprofits. When a hospital is finally sold, the officers and directors do not split up the proceeds. The nondistribution clause imposed by the IRS ensures that.

So, in a real sense, what we need to do is to shift our focus from "ownership" to "control." The asset value of a building is realized only when it is sold, and then the proceeds are not distributed. Depreciation is essentially meaningless, since it has no tax benefits for tax-exempt organizations. We know that our accountant is not going to let us defame depreciation, but truthfully, how many nonprofit organizations are prepared to establish plant funds to compensate for it? (Plant funds are reserves designed to provide money for replacement and renovation of facilities.) Few nonprofits can afford to sustain large liquid plant funds, and even if they could, the costs of renovation are not deductible.

At one time or another, all of us have dealt with a non-profit which owned a building that it could not afford to operate, or that it could not maintain in excellent condition. Serving two masters—social purpose and fiscal solvency—makes it difficult to allocate resources to a plant fund when the orphans are hungry.

■ *What if someone wants to give us a building?*

Accepting gifts always seems like a wonderful idea. After all, we all love presents. If someone wants or needs to give you a building, you can accept with gratitude, if you have made provision for someone else to own and operate it. Why not arrange the donation and then sell the equity in the building over to a businessperson. Because you wish to control the space, find a person who will accept the value of the building as prepaid rent and utilities for a period of time. You get to control the space on a long-term lease, the donor gets to deduct the value of the building from his or her income taxes, and the businessperson gets the opportunity to enjoy the tax benefits of ownership.

■ *We already have a building. Why wouldn't we want to keep it?*

You can. But you might want to consider—just consider, mind you—the possibility that your institution might be better off if you sold the building to an interested party, secured a really long-term lease that gave you lots of control, and used a significant portion of the proceeds from the sale to fund your rental for many, many years. Then, when the roof fails in the next snowstorm, someone else's insurance can cover the costs, and someone else who can derive a tax benefit can take care of the consequences. While all this is happening, you can continue providing a public good.

■ *But isn't our building giving us equity for the future?*

If your nonprofit were a business and had equity to offer, then the building would serve that purpose. But because your organization is a nonprofit, all the building has is an asset value. You can sell the building, but you will not distribute the proceeds; instead they will be used to lease or buy or build another space, or they may be converted into program funds to help fulfill your mission.

■ *Shouldn't we consider real estate as an investment opportunity?*

56

Ownership is often justified by assuming that property can be acquired and held as an investment that will yield future profits. Nonprofit organizations are prevented from establishing an equity position with regard to assets. For this reason, social purposes should have first claim on all donated funds. Buildings do not lower program costs, and the funds used to acquire buildings are no longer available to pursue the institution's mission.

■ *Our donors will not support operations, but they will fund buildings. What can we do?*

Some nonprofits use the "edifice complex" to rationalize acquiring real estate. They claim that contributors are willing to donate for bricks and mortar, but not for less "concrete" purposes such as research or operating expenses.

Many institutions, such as cultural organizations, colleges and universities, and hospitals, find it easier at times to raise money for buildings than for equally worthy but less permanent and visible purposes, such as scholarships or operating funds.

However, often the edifice complex is actually encouraged by the "conventional wisdom" of development officers and fund-raising consultants, not the donors. When nonprofits analyze the costs and benefits of owning versus leasing, the financial gains of leasing are usually compelling. These numbers, presented in conjunction with the costs of conducting a major capital campaign, are often attractive to donors with business acumen.

■ *Are you urging us to lease?*

We advise our nonprofit clients to try not to own things that are not disposable. Leasing is possible in many different forms, and concessionary leasing is often less expensive than commercial equipment rental. The reason for not owning is simply that there is no incentive in the nonprofit world to own. There are no tax advantages for tax-exempt organizations. Depreciation takes place, but your institution funds it at the cost of fulfilling its social purpose. When things wear out or need to be replaced, it makes more sense to have people who can derive benefit from investing do so, rather than burdening the organization and the people you serve with increased costs.

■ *This all sounds like a lot of work. Isn't it easier to just get the money up front, and then buy or build a building?*

For an institution contemplating a capital campaign, the first consideration is the cost of raising money. There is considerable speculation within the nonprofit sector about the actual costs of raising funds. One organization that has clearly identified these costs is the Association of Health Care Philanthropies, which has studied the issue for years.

Among their member institutions, those that have been raising significant sums of money for more than 10 years spend a median of 33 cents to raise a dollar, while those that have been fund-raising for less than two years spend a median of 77 cents to raise a dollar. The major variable in this survey was time.

Successful capital campaigns must include some additional funding to account for donor fatigue after the goal is achieved. Often campaigns also include an endowment to provide operating funds in the future. When these costs are weighed against rental expenses, there is usually a considerable discrepancy. Mobilizing all that money up front can be very expensive and time consuming.

■ *Isn't it cheaper to own than to rent?*

Many people believe that leasing facilities is ultimately more expensive than owning them. This impression is often reinforced by unrealistic sentiments regarding ownership. Often ignored are the expenditures necessary to conduct a successful capital campaign. In addition, if future operating funds are placed in an endowment, this capital is being inefficiently used. Finally, the costs of future maintenance on buildings are often minimized. Many people shy away from leasing because they believe that when an organization rents a building, it is charged an amount that is intended to cover the landlord's property taxes, interest fees, insurance costs, and profit. Avoiding leasing for this reason ignores the basic economic laws of supply and demand.

The rent or lease payments received by a building owner are determined not by the cost of the building, the interest charges, or the insurance fees. Property owners do not set rents on a cost-plus basis that guarantees them a profit; rather, they charge what the market will bear.

■ *Since every state is different, how do you know that these principles apply in my area?*

Real estate markets are different from state to state. Proponents of ownership usually claim that renting is cheaper in some states and ownership is less costly in others. But this explanation is more contrived than convincing. There is no evidence that the laws of supply and demand that govern the real estate market differ significantly across state lines. The value or price of a commercial building is determined by the annual income generated by rent or lease payments. So when rents are low, buildings sell at low prices; conversely, when rents are high, buildings command high prices. Rents and purchase prices move together; thus a building purchased "cheaply" can also be rented "cheaply."

Yet many nonprofit managers would have donors believe that when rents are dear, they can purchase buildings cheaply. If this is the case, then those managers are definitely in the wrong business—there are vast fortunes to be made in real estate.

■ *Control is important. If we own our building, we have control. How can we control our space if we are leasing?*

Ownership and control are not necessarily the same thing. Owning property that you can't afford to maintain can have disastrous effects on your program budget.

Control, on the other hand, can be secured through well-crafted long-term leases that bind the parties to a set of understandings that are workable and fair. By spelling out terms and conditions in advance, the nonprofit organization can plan for its future expenditures without worrying about unanticipated disasters.

■ *Do leasing arrangements for nonprofit organizations provide opportunity to investors?*

Ample opportunity exists for socially motivated investors to participate in creative leasing arrangements with nonprofit organizations. In many cases, an institution that has a projected revenue stream can benefit from investment funds offered on concessionary terms. Socially motivated investors or a foundation with program-related investments can participate in the leasing arrangement. Through the use of tax incentives, innovative partnerships with banks, special government programs, and individual investors, leases can be structured to provide benefits to all parties.

After the numbers on leasing have been run, it makes sense to develop a business plan aimed at local partners, potential investors, and affiliated agencies and institutions. The plan should include a long-range forecast that demonstrates the institution's ability to generate income to pay its operating expenses and, particularly, the rent. Investors motivated to participate may bring a sense of social purpose to the initial meeting, but they will need to be convinced that the investment is solid before proceeding. It is sometimes helpful to develop before-and-after tax scenarios for potential investors.

In some instances, a financing plan developed in cooperation with a local bank makes a compelling case. If you can arrange for the financing to be assumed by investors, some risk to the bank is mitigated and additional credibility is offered to the deal.

There are a number of funding institutions that can make concessionary rates available to organizations with ongoing cash flow. These sources may also enhance the position of the lender or the investor. In some cases, tax-free bonds may be available from a local municipality or agency, which can reduce the costs of interest and make the investment more attractive.

The goal is to ensure that the nonprofit organization is able to secure long-term control of space, maximize the use of its own capital, and reduce its immediate and long-term expenses through the payment of rent to an owner who can absorb the primary advantages of ownership. In this way, the social purpose of the nonprofit organization is preserved as the focus for community support.

Earned Revenue

Maybe earning more money will solve our problem. Don't we just need to be more entrepreneurial?

After the grant-making wave of the '70s, a new interest in earned revenue swept across the nonprofit sector. All sorts of schemes were proposed and adopted. Many of them generated income, but often at great cost to the organizations. If we sneak a peek at some of the premises underlying the earned revenue revolution, it may help us to understand why things didn't always turn out to be better in entrepreneurial terms.

The vast majority of large businesses in American society operate with a relatively small margin of profit. Supermarkets, for example, spend 99 cents to raise one dollar. Margins of 5%, 6%, or 7% are not uncommon in the ranks of many competitive corporations. Do nonprofits, already strapped for cash, have the capital to invest in running a store that will bring in a surplus of less than 10%? Many do not, and those that do often fail to grasp how time consuming and demanding successful business undertakings can be. Will the time and effort devoted to obtaining revenue be a drain on the organization and the social purpose for which the organization was chartered? In other words, when does the quest for income overwhelm the mission of the institution?

In boardrooms across the country, earned revenue strategies are hatched and fueled by the idea that nonprofits can be more efficient than businesses. Yet, if we look at a direct comparison between tax-exempt credit unions and banks, we see that the former are not more profitable than the latter, despite being substantially free from the burden of taxation. The underlying premise that nonprofits are more efficient than their commercial counterparts is questionable.

What about the role of volunteers? They admittedly reduce some costs, but also add costs through turnover and the need for flexible scheduling and efficient coordination. In the short run, volunteerism may be great, but in an entrepreneurial environment, in the long run it is not always more effective than a work force.

When you attend a production at a local nonprofit theater company, and the program has a note telling you that your ticket has paid for half of the cost of this evening's production, you might translate this statement to read: "We just spent two dollars to earn a buck from you." Without help from corporations, foundations, governmental agencies, and individual donors to make up that additional dollar, the house would be dark.

Discounts

■ *We offer discount sales to season ticket holders. Isn't this a good way to make money? And if we don't, how are we going to survive?*

Subscription campaigns have a short and checkered history. They were conceived as a way to fill seats and the organization's coffers to cover seasonal deficits and front new productions. Instead, they have proven to be costly and, in many ways, questionable as a financial strategy.

Here's why. When an organization offers a discount of, say, 15% to subscribers, it is taxing itself at a higher rate of interest than a bank might charge to lend. Added to the discount are the costs of the subscription campaign, which can be as much as 5% of the annual budget. Having two campaigns, one for fund-raising and the other for subscribers, frequently confuses both donors and subscribers.

One evening, we got a call at dinner from a local theater asking us to subscribe and offering a 15% discount to do so. I assumed that this campaign probably cost at least 5% to conduct, so the net cost to the theater was 20%. Less than an hour later, we got a call from another telemarketing firm, working for the same theater, telling us how much our donation was needed to cover costs. We did a little math on the phone, and, sure enough, the difference was close to 20%. I suggested that the theater not offer financial discounts, but the person on the other end of the phone wanted a pledge, not free advice.

The next day, I called the manager of the theater and made an appointment to discuss this interesting paradox. During our meeting, I asked how the theater's board and staff had arrived at a 15% discount. He said the number had sounded like it would be a good inducement to prospective subscribers.

At that point in the discussion, I explained that I had, some years earlier, surveyed subscribers in a city-wide assessment of subscription campaigns and found that almost none of them were greatly influenced by price. In fact, the vast majority told me that they supported their cultural institutions and liked to subscribe in order to get preferential seating or greater flexibility in ticket exchange, but would be just as happy to get a non-financial perk such as good coffee and a cookie at intermission, or more flexible ticket exchange, or invitations to special events. I suggested that the theater check with their own subscribers to confirm or deny my statements, but that our research indicated that the people who are most price sensitive are the single-ticket purchasers. Yet most theaters charge them the highest prices and offer them the least benefits.

"Drop the financial discounts," I said, "and serve them Seattle-roasted coffee and cookies instead." The theater manager eyed me coldly and said, "Just how do you propose that I find the money I need to pay off last year's deficit and to front this year's productions?" I answered, "Look, form a credit holders group that will support your mission; then go to the bank and engage in a little secured borrowing. See if your budget and balance sheet don't look better in a year. Get your money at three points below prime, rather than paying the equivalent of 20% in interest through all those discounts and campaign costs."

Together we did the math, and the figures were impressive. The theater manager said he would try it, and as I was leaving I suggested that he drop single-ticket prices for a while to bring in new audience members and see if attendance increased. A year later, the theater was functioning much better, having effected each of these changes, and they called us only once to solicit funds— of course, still in the middle of supper.

Conclusion

When all is said and done, overseeing the finances of a nonprofit organization can be enjoyable, rewarding, and even interesting for you in your role as a board member. See it as a progression that starts with cash flow budgeting and leads to honest forecasting and diligent tracking of income and expense.

Ask that the financial reports be reasonable, direct, and understandable. After all, it's your job to grasp them.

Establish credit early on with your bank, and use it wisely to stabilize your cash flow. It can bring a variety of benefits.

Avoid activities that do not contribute efficiently to cash flow, such as endowments or cash reserves, or that are not effective for the organization in the long run, such as building or equipment ownership.

Finally, try not to accumulate capital. Instead, spend your funds on your mission, and use your credit to insure your future. It really is simple, and, after all, money matters in nonprofit organizations.

References

Anthes, E., Cronin, J., and Jackson, M., *The Nonprofit Board Book*, Independent Community Consultants, 1985.

Baumol, William J., and William Bowen, *Performing Arts: The Economic Dilemma*, MIT Press, 1966.

Beatty, J., "The Middle-Class Crisis," *Atlantic Monthly*, 1994.

Bennett, J.T., and DiLorenzo, T.J., *Unhealthy Charities*, Basic Books, 1994.

Blake, M., "A Study of Capital Giving to San Francisco Nonprofits, 1988-1992," The Water & Elise Haas Fund, 1994.

Bolman, L.G., and Deal, T.E., *Reframing Organizations: Artistry, Choice, and Leadership*. Jossey-Bass 1994.

DiMaggo, P.J. *Nonprofit Enterprises in the Arts*, Oxford Press, 1986.

"What's Reasonable? Investment Expectations and the Foundation Pay-Out Rate," *Foundation News*, January/February 1976.

Hammack, D.C., and Young, D.R., *Nonprofit Organizations in the Market Economy*, Jossey-Bass, 1993.

Hopkins, B.R., *The Legal Answer Book for Nonprofit Organizations*, John Wiley & Sons, New York, 1992.

Kauffman, S.A., *Origins of Order: Self-Organization and Selection in Evolution*, Oxford University Press, 1992.

Linzer, R., "Borrowing: A Resource for Nonprofits," *Chronicle of Non-Profit Enterprise*, January 1989.

Linzer, R., "Why You Want to Borrow from Banks," *Taft Nonprofit Executive*, Volume VIII, Number 8, April 1989.

Linzer, R., "Endowments: The Other Side of the Coin," *Chronicle of Non-Profit Enterprise,* July 1992.

Linzer, R., "Creditholders," *The Practical Philanthropist*, July 1992.

Linzer, R., "Using Credit to Stabilize Cash Flow," *Land Trust Alliance Exchange*, Winter 1993.

Linzer, R., "Nonprofits: National Trends and Issues," speech prepared for cultural institutions in Reno, Nevada, February 27, 1996.

Linzer, R., "Charities Should Borrow Money, Not Hoard It," Opinion Section, *The Chronicle of Philanthropy*, July 24, 1997.

Rudney, G., "The Scope and Dimensions of Nonprofit Activity," *The Nonprofit Sector: A Research Handbook*, Yale University Press, 1987.

Skloot, E., *Smart Borrowing: A Nonprofit's Guide to Working with Banks*, New York Community Trust, 1989.

Stevens, S. and Lisa Anderson, *All the Way to the Bank: Smart Money Management for Tomorrow's Nonprofit*, The Stevens Group, St. Paul, Minnesota, 1997.

Waldrop, M.M., *Complexity: The Emerging Science at the Edge of Order and Chaos*, Touchstone, 1992.

Williamson, P.J., *Spending Policy for College and University Endowments*, The Common Fund, Fairfield, Connecticut, 1979.

Richard and Anna Linzer

Facilitation + Consultation

Since 1965, Richard Linzer has provided consultation for businesses, nonprofit organizations, and government agencies. He has worked with organizations in the areas of financial management, board development, institutional analysis, and strategic planning.

Richard Linzer specializes in the use of credit by nonprofit organizations. His articles on financial management have appeared in the *Chronicle of Non-Profit Enterprise, Taft Nonprofit Executive, The Practical Philanthropist, Land Trust Alliance Exchange,* and *The Chronicle of Philanthropy.* He has taught seminars on financial management for non-profit organizations and is frequently a speaker on this topic at conferences.

Anna Linzer is a poet and writer. Her collection of stories is *Ghost Dancing,* published by Picador of St. Martin's Press, 1998. Her poems and short stories have appeared in literary magazines and anthologies, including *Kenyon Review, Carolina Quarterly,* and *Red Earth, Blue Dawn.*

In addition, Richard and Anna work together as facilitators for meetings and retreats. They have collaborated to create workbooks and kits to help groups work more effectively and efficiently. They are co-authors of *The Board Development Kit, The Board Retreat Kit, The Corporate Retreat Kit, The Collaboration Workbook,* and *The Strategic Planning Kit for Public Agencies.*